I. Introduction

This guidance provides technical recommendations for protecting underground sources of drinking water (USDWs) from potential endangerment posed by hydraulic fracturing (HF) activities where diesel fuels are used.

The U.S. Environmental Protection Agency (EPA) developed this guidance for EPA permit writers to ensure protection of USDWs in accordance with the Safe Drinking Water Act (SDWA) and Underground Injection Control (UIC) regulatory authority. This authority is limited to when diesel fuels are used in fluids or propping agents pursuant to oil, gas and geothermal activities. This document does not establish any new permitting requirements for HF activities using diesel fuels, but describes the EPA's interpretation of existing legal requirements as well as non-binding recommendations for EPA permit writers to consider in applying UIC Class II[1] regulations to HF when diesel fuels are used in fracturing fluids or propping agents. This document does not address geothermal activities.

The EPA expects that EPA UIC Program Directors, and the permit writers acting on their behalf, will follow the interpretation of the statutory term "diesel fuels" presented in this guidance document. They should also consider, although are not required to follow, the recommendations reflected in this guidance on how to apply the Class II regulations to HF activities using diesel fuels when issuing permits for such activities under the federal UIC Program. Recommendations are consistent with the discretion accorded under the existing UIC Class II regulations, and reflect existing UIC requirements for other well classes, voluntary industry standards, state rules, and other model guidelines for HF. However, permit writers, acting on behalf of the UIC Director have the discretion to consider alternative approaches that are consistent with statutory and regulatory requirements. Decisions about permitting HF operations that use diesel fuels will be made on a case-by-case basis, considering the facts and circumstances of the specific injection activity and applicable statutes, regulations and case law.

Under the 2005 amendments to the SDWA, a UIC Class II permit must be obtained prior to conducting the underground injection of diesel fuels for hydraulic fracturing. The EPA, where it directly implements the program, as well as states and tribes with primary enforcement authority, must issue a Class II permit prior to the injection of diesel fuels in the HF fluid or propping agents. The primary audience for these technical recommendations is the EPA Regional offices directly implementing the existing UIC Class II Program requirements (40 *Code of Federal Regulations* (CFR) parts 124 and 144 through 147).

Stakeholders and the public have recognized the importance of safely and responsibly managing unconventional oil and gas development, including hydraulic fracturing. Many states have updated their oil and gas regulations and a variety of organizations have developed model guidelines and best practices. The EPA engaged with states, tribes, industry, and other stakeholders during the development of this document and reviewed best practices available at

---

[1] Class II is the primary well classification used for injection wells that are associated with oil and gas storage and production (40 CFR 144.6).

the time. The EPA used information from these efforts to inform this guidance for the UIC program.

An EPA analysis of data on HF fluids posted in 2012 on the chemical disclosure registry website FracFocus[2] found that diesel fuels appeared in fewer than two percent of the wells.[3] While FracFocus data are voluntarily submitted and not statistically representative of the presence of diesel fuels or other chemical substances in HF fluids, they are useful in providing an indication of the extent to which industry relies on diesel fuels for HF activities. While diesel fuels as defined in this guidance are currently used in a small percentage of HF wells, the EPA will work with states and industry to promote best practices in HF operations, including partnering with stakeholders to support voluntary use of greener alternatives in HF fluids generally.

Although developed specifically for hydraulic fracturing where diesel fuels are used, many of the recommended practices found in this document are consistent with best practices for hydraulic fracturing in general, including those found in state regulations, voluntary standards from the American Petroleum Institute (API), and model guidelines for hydraulic fracturing developed by industry and stakeholders. In particular, the EPA's recommendations for applying UIC requirements on area of review, well construction, operations, and monitoring – including testing for mechanical integrity of the well and baseline and follow-up water quality monitoring – will also promote adoption of some best practices identified by industry, states, and other groups.

The practices described in this guidance are critical for ensuring that underground sources of drinking water are protected during hydraulic fracturing using diesel fuels. For example, delineating a site-specific area of review, including for the horizontal section of a well, ensures there are no conduits that could allow the escape of contaminants into USDWs. During the area of review delineation an owner/operator looks for artificial or natural conduits to ensure adequate confinement and takes corrective action if necessary to prevent fluid or gas migration. Similarly, mechanical integrity tests (MITs) ensure that the protective physical components of the well, including the casing and cement, are competent prior to injection and throughout the life of the well. High injection pressures, such as those occurring during HF, have the potential to damage the mechanical integrity of the well causing leaks, which may allow for the migration of fluids into USDWs. Conducting MITs ensures that injection well integrity is maintained at all times. Baseline and post-fracture water quality monitoring are used to help ensure that a permitted well has not endangered USDWs.

In addition to reflecting UIC program requirements, state regulations and industry best practices, a number of the practices contained in this guidance were outlined by the Secretary of Energy's Advisory Board (SEAB) Shale Gas Production Subcommittee in its August and November 2011 reports (US DOE, 2011). Thus, states and tribes responsible for issuing permits and/or updating regulations for hydraulic fracturing may find the recommendations in this document useful in improving the protection of underground sources of drinking water and public health wherever hydraulic fracturing is practiced.

---

[2] FracFocus (http://fracfocus.org/) is the national HF chemical registry managed by the Ground Water Protection Council and Interstate Oil and Gas Compact Commission.
[3] An August 2012 search of FracFocus identified only one well that used diesel fuels as a carrier fluid.

II. Background Information

How are diesel fuels used in the HF operations?

HF is a technique used to produce economically viable quantities of oil and natural gas, especially from unconventional reservoirs, such as shale, tight sands, coalbeds and other formations. HF involves the injection of fluids (commonly a mixture of water, chemical additives and proppants) under pressures great enough to open and enlarge fractures within the oil-and gas-producing formations. The resulting fractures are held open using propping agents, such as fine grains of sand or ceramic beads, to allow oil and gas to flow to the production well. The types and concentrations of chemical additives and proppants used in the HF fluids vary depending on site-specific conditions and are usually tailored to the properties of the formation and the needs of the project.

Diesel fuels are among a number of oil-based fracturing fluids that can be used to avoid damage such as reduced permeabilities to water sensitive formations and allow for better production (DeVine et al., 2003). Diesel fuels may be used as an additive to adjust fluid properties (e.g., viscosity and lubricity) or act as a solvent to aid in the delivery of gelling agents. Diesel fuels' properties of high viscosity and immiscibility in water may also prevent fluid leak-off or loss into a formation without impeding the production of hydrocarbons (McCabe et al., 1990; Rae and DiLullo, 1996). Also, the lower freezing point of diesel fuels relative to water may be useful in cold climate operations as an effective winterizing agent by preventing liquids from freezing in low temperatures (Shibley and Leonard, 1987).

Diesel fuels may contain a number of chemicals of concern including benzene, toluene, ethylbenzene, and xylene compounds (BTEX). BTEX compounds are highly mobile in ground water and are regulated under the SDWA national primary drinking water regulations (NPDWRs) because of the risks they pose to human health. The EPA has set a maximum contaminant level goal (MCLG)[4] and a maximum contaminant level (MCL)[5] for each compound. For example, the MCLG for benzene is zero and the MCL is 0.005 mg/L.[6] People consuming drinking water containing any of these chemicals in excess of the standards set by the EPA over many years could experience:

- An increase in anemia or a decrease in blood platelets from benzene exposure[7];
- An increased risk of cancer from benzene exposure[8];
- Problems with the nervous system, kidneys or liver from toluene exposure[9];
- Problems with the liver or kidneys from ethylbenzene exposure[10]; and

---

[4] The EPA sets the level of protection for MCLGs based on the best available science to prevent potential health problems.
[5] The EPA sets MCLs as close to the MCLGs as possible, considering cost, benefits and the ability of public water systems to detect and remove contaminants using suitable treatment technologies.
[6] US EPA, http://water.epa.gov/drink/contaminants/basicinformation/benzene.cfm
[7] US EPA, *Ibid*
[8] US EPA, *Ibid*
[9] US EPA, http://water.epa.gov/drink/contaminants/basicinformation/toluene.cfm

- Damage to the nervous system from exposure to xylene[11].

BTEX compounds are classified as aromatic hydrocarbons, a class of substances found in petroleum products including diesel fuels. The total content of aromatic hydrocarbons in petroleum products varies based on the refining process. The diesel fuels identified in this guidance memorandum can contain up to 25 percent aromatic hydrocarbons, by weight (API, 2012). These diesel fuels can also contain 20 to 60 percent polynuclear aromatic hydrocarbons (PAHs) by volume (API, *Ibid*). PAHs can be a toxic component of petroleum products and some PAHs are listed as Priority Pollutants under the Clean Water Act[12].

Because other substances used in HF fluids may contain similar levels of BTEX, even if the Chemical Abstract Service Registry Number (CASRN) does not identify the substance as diesel fuel, the EPA will work with states and industry to explore approaches to promote voluntary use of safer alternatives in HF fluids.

The EPA conducted an analysis of data on HF fluids posted in 2012 on the chemical disclosure registry website FracFocus to determine how diesel fuels are currently used in HF operations. Based on this analysis, diesel fuels were most commonly used as an additive to reduce friction. Diesel fuels appeared in fewer than two percent of the wells,[13] and no regional patterns of diesel fuels usage were identified from data registered in FracFocus.

When does a HF activity require a UIC Class II permit?

A HF activity is subject to UIC Class II permitting requirements under the SDWA if any portion of the injectate contains "diesel fuels." The EPA interprets this statutory term to mean any of the following five CASRNs:

- **68334-30-5 Primary Name: Fuels, diesel**
  Common Synonyms: Automotive diesel oil; Diesel fuel; Diesel oil (petroleum); Diesel oils; Diesel test fuel; Diesel fuels; Diesel fuel No. 1; Diesel fuel [United Nations-North America (UN/NA) number 1993]; Diesel fuel oil; European Inventory of Existing Commercial Chemical Substances (EINECS) 269-822-7.

- **68476-34-6 Primary Name: Fuels, diesel, No. 2**
  Common Synonyms: Diesel fuel No. 2; Diesel fuels No. 2; EINECS 270-676-1; No. 2 Diesel fuel.

- **68476-30-2 Primary Name: Fuel oil No. 2**
  Common Synonyms: Diesel fuel; Gas oil or diesel fuel or heating oil, light [UN1202] No. 2 Home heating oils; API No. 2 fuel oil; EINECS 270-671-4; Fuel oil No. 2; Home heating oil No. 2; No. 2 burner fuel; Distillate fuel oils, light; Fuel No. 2; Fuel oil (No.

---

[10] US EPA, http://water.epa.gov/drink/contaminants/basicinformation/ethylbenzene.cfm
[11] US EPA, http://water.epa.gov/drink/contaminants/basicinformation/xylenes.cfm
[12] 40 CFR Part 423 (Appendix A)—126 Priority Pollutants
[13] See footnote 3.

1,2,4,5 or 6) [NA1993].

- **68476-31-3 Primary Name: Fuel oil, No. 4**
  <u>Common Synonyms</u>: Caswell No.[14] 333AB; Cat cracker feed stock; EINECS 270-673-5; EPA Pesticide Chemical Code 063514; Fuel oil No. 4; Diesel fuel No. 4.

- **8008-20-6 Primary Name: Kerosene**
  <u>Common Synonyms</u>: JP-5 navy fuel/marine diesel fuel; Deodorized kerosene; JP5 Jet fuel; AF 100 (pesticide); Caswell No. 517; EINECS 232-366-4; EPA Pesticide Chemical Code 063501; Fuel oil No. 1; Fuels, kerosine; Shell 140; Shellsol 2046; Distillate fuel oils, light; Kerosene, straight run; Kerosine, (petroleum); Several Others.

The use of diesel fuels in oil and gas production applications is not subject to UIC Class II permitting requirements in certain cases. Specifically, those cases are non HF activities such as when diesel fuels are a component of drilling muds or pipe joint compounds used in the well construction process, or when diesel fuels are used in motorized equipment at the surface.

---

[14] A Caswell No. is an alphanumeric chemical identifier implemented by Robert L. Caswell in the 1960s and 1970s in conjunction with acceptable common names of pesticides for labeling purposes.

III. Technical UIC Program Requirements and Recommendations for Application to Hydraulic Fracturing Activities Using Diesel Fuels

This section of the guidance addresses the questions listed below. For each question, the document provides a brief summary of the existing federal UIC Class II Program regulations followed by the EPA's recommended approaches for EPA Regional offices to consider when permitting the use of diesel fuels during HF. This section is not intended to present UIC Class II permit requirements in their entirety. Readers seeking more information about Class II permit requirements should refer to 40 CFR part 124 and parts 144 through 147.

EPA UIC Program Directors, and the permit writers acting on their behalf, should consider the recommendations reflected in this section when issuing permits applying the Class II regulations for HF activities using diesel fuels under the federal UIC Program. Recommendations are consistent with the discretion accorded under the existing UIC Class II regulations, and reflect existing UIC requirements for other well classes, voluntary industry standards, state rules, and other model guidelines for hydraulic fracturing. However, permit writers, acting on behalf of the UIC Director have the discretion to consider alternative approaches that are consistent with statutory and regulatory requirements. Decisions about permitting HF operations that use diesel fuels will be made on a case-by-case basis, considering the facts and circumstances of the specific injection activity and applicable statutes, regulations and case law.

The questions addressed by this guidance are as follows:

- What Are Considerations in the Submission and Review Process for Diesel Fuels HF Permit Applications?

- What Information Should Be Submitted with the Permit Application?

- Can Multiple UIC Class II Wells Using Diesel Fuels for HF Be Authorized by One Permit?

- How Should EPA UIC Permit Writers Establish Permit Duration and Apply UIC Class II Requirements After HF at a Well Ceases?

- How Do the Area of Review (AoR) Requirements at 40 CFR 146.6 Apply to Wells Using Diesel Fuels for HF?

- How Do the Class II Well Construction Requirements Apply to HF Wells Using Diesel Fuels?

- How Do the Class II Well Construction Requirements Apply to Already Constructed Wells Using Diesel Fuels for HF?

- How Do the Class II Well Operation, Mechanical Integrity, Monitoring, and Reporting Requirements Apply to HF Wells Using Diesel Fuels?

- How Do the Class II Financial Responsibility Requirements Apply to Wells Using Diesel Fuels for HF?

- What Public Notification Requirements or Special Environmental Justice (EJ) Considerations are Recommended for Authorization of Wells Using Diesel Fuels for HF?

What Are Considerations in the Submission and Review Process for Diesel Fuels HF Permit Applications?

Existing Class II Requirements in 40 CFR parts 124 and 144 through 147:

For the purposes of UIC Class II permitting, any well injecting diesel fuels for HF is considered a "new injection well" (40 CFR 144.31), even if it was originally constructed as an oil and gas production well and requires a UIC Class II permit (40 CFR part 124 and parts 144 through 147). Permits for diesel fuels HF are required to be approved prior to commencing all injection-related activities, including injection, well construction, retrofitting components of an existing well and commencing the HF process. An owner or operator seeking a UIC permit for injection must submit an application for a permit as expeditiously as practicable and in a reasonable amount of time prior to the expected start of construction or injection, as determined by the UIC permit writer (40 CFR 144.31).

Recommendations for applying existing requirements to HF activities using diesel fuels:

The paragraphs below present the EPA's recommended approaches for applying the existing Class II requirements described above to the permit application submission and review process for wells where diesel fuels will be used during HF. Recommendations are consistent with the discretion accorded under the existing UIC Class II regulations, and reflect existing UIC requirements for other well classes, voluntary industry standards, state rules, and other model guidelines for hydraulic fracturing.

EPA UIC Program Directors should establish a process for the timely submission and review of permit applications consistent with 40 CFR 144.31 that allows sufficient time to review and authorize the permit prior to initiating HF activities using diesel fuels. The application timeframe should allow time to evaluate the proposed use of diesel fuels for HF to ensure that injection will not endanger USDWs. The permit review timeframe should be of a sufficient duration to allow the EPA UIC permit writer to comprehensively consider all relevant permit information, such as proposed construction, operation and monitoring plans, to establish appropriate permit conditions and to include an opportunity for public notice and comment prior to issuing approval of the UIC Class II permit for wells using diesel fuels for HF. The EPA will provide tools, such as checklists, to help owners or operators develop complete permit applications in order to increase the likelihood of timely review.

EPA UIC Program Directors should continue to coordinate with state oil and gas programs and the appropriate Bureau of Land Management (BLM) office to establish a mechanism to inform owners or operators of applicable UIC Program requirements and application deadlines. Multiple mechanisms for outreach should be used to notify owners or operators of expected permit application review and approval timeframes thereby preventing delays for drilling and construction.

Collaboration among regulatory entities is important so that appropriate parties are aware of situations where owners or operators plan to use diesel fuels for HF. For example, all parties can work together to streamline permitting (e.g., between the EPA and BLM on BLM-managed lands or with state agencies) such as sharing data where compliance requirements and reporting timeframes are sufficiently compatible for coordination among the various permitting authorities. Regional EPA UIC Class II Programs should reach out to their state oil and gas programs to determine collaborative ways to notify potential owners or operators early regarding the various permitting requirements that may apply. For example, a check box, notation or UIC Program contact information can be added to the oil and gas drilling permit application checklist to alert owners or operators using diesel fuels for HF of the need to apply for a Class II UIC permit.

What Information Should Be Submitted with the Permit Application?

Because of the high injection pressures, the potential to induce fractures that may serve as conduits for fluid migration, and the particular risks associated with diesel fuels, EPA UIC permit writers must evaluate a variety of factors in reviewing the permit application to ensure that appropriate safeguards (e.g., permit conditions) are established during the permitting process to prevent potential contamination of USDWs.

Existing requirements in 40 CFR parts 124 and 144 through 147:

Existing UIC Class II Program provisions at 40 CFR parts 124 and 144 through 147 require owners or operators to submit information to the UIC Program Director to consider before issuing any Class II permits. Select submission requirements from the existing Class II regulations are listed below:

- Maps showing the injection well or project area for which the permit is sought and the applicable area of review (AoR) showing the number or name and location of all producing wells, injection wells, abandoned wells and other features (40 CFR 146.24(a)(2));

- All known wells within the AoR[15] or zone of endangering influence (ZEI) that penetrate formations affected by the increase in pressure (40 CFR 146.24(a)(3));

---

[15] See the section entitled "How Do the Area of Review (AoR) Requirements at 40 CFR 146.6 Apply to Wells Using Diesel Fuels for HF?" for additional information.

- Data[16] on the injection and confining zones including lithologic description, geological name, thickness and depth and estimated fracture pressures of the injection and confining zones (40 CFR 146.24(a)(5));

- The location, orientation and properties of known or suspected faults and fractures that may transect the confining zone(s) in the AoR and a determination that they would not interfere with containment (40 CFR 146.24(a)(2));

- Geologic name and depth to the bottom of all USDWs, which may be affected by the injection (40 CFR 146.24(a)(6));

- Well construction schematics including surface and subsurface details (40 CFR 146.24(a)(7));

- Proposed stimulation (fracturing) program (40 CFR 146.24(b)(2)) and the proposed injection procedure for each stage of the HF (40 CFR 146.24(b)(3));

- Operating data, such as average and maximum daily rate, volume, and injection pressure of fluids to be injected. The source and an appropriate analysis of the chemical and physical characteristics of the injection fluid (40 CFR 146.24(a)(4));

- A detailed chemical plan describing the proposed HF fluid composition,[17] and an appropriate analysis of the chemical and physical characteristics of the HF fluid, including the volume and range of concentrations for each constituent (40 CFR 146.24(a)(4)(iii));

- Names and addresses of all owners of record of land within one-quarter (¼) mile of the well boundary (40 CFR 144.31(e)(9)). In the case of diesel fuels HF this includes the names and addresses of all owners of record of land within a (¼) mile fixed radius around the wellhead, facility boundary or within the boundaries of the ZEI.

- Appropriate logs and other tests conducted during the drilling and construction of wells and reports interpreting the results of the tests as described in 40 CFR 146.24(c)(1); and

- A plugging and abandonment plan that meets the requirements of 40 CFR 146.10, which describes the need to cement a well to prevent fluid movement (40 CFR 144.31(e)(10)).

Recommendations for applying existing UIC requirements to HF activities using diesel fuels:

The section below presents recommendations for applying the existing requirements described above regarding information that should be submitted with a permit application for wells where

---

[16] Data may include geo-mechanical characteristics such as: fracture stress, ductility, rock strength, in situ fluid pressures and others.

[17] Owners or operators may make claims of confidentiality regarding this information (40 CFR 144.5).

diesel fuels will be used during HF. Recommendations are consistent with the discretion accorded under the existing UIC Class II regulations, and reflect existing UIC requirements for other well classes, voluntary industry standards, state rules, and other model guidelines for hydraulic fracturing.

EPA UIC Program Directors should consider requesting the following information from the owner or operator, per their authorized discretion under 40 CFR 144.52(a)(9):

- Information about the extent and orientation of the planned fracture network, any nearby USDWs and their connections to surface waters, if any,[18] as well as any other information that can be used to understand, calculate and delineate the extent and orientation of the fracture system expected to be created by the proposed diesel fuels HF activity. This includes results from previous HF operations in the area and other empirical information, models and published studies and reports;

- In situations where permits include a duration that is shorter than the full life of the well, a pre-permit-expiration monitoring plan that incorporates water quality monitoring in the AoR may be needed to demonstrate non-endangerment. Monitoring parameters could include ground water flow and depth; total dissolved solids (TDS); specific conductance; pH; chlorides; bromides; acidity; alkalinity; sulfate; iron; calcium; sodium; magnesium; potassium; bicarbonate; detergents; diesel range organics (DRO); benzene, toluene, ethylbenzene and xylenes (BTEX); isotopic methane and/or radionuclides (40 CFR 144.51(j) and 40 CFR 146.24(a)(4)(iii));

- Information on seismic history, such as the presence and depth of known seismic events in areas where prior seismic activity would lead the UIC Program Director to be concerned about endangerment of USDWs (40 CFR 146.24(a)(2));

- Baseline geochemical information on accessible USDWs and other subsurface formations of interest within the AoR of a Class II diesel fuels HF well (40 CFR 146.22(b)(2)(i) and (f)(2)).[19] This geochemical information could include parameters such as TDS; specific conductance; pH; chlorides; bromides; acidity; alkalinity; sulfate; iron; calcium; sodium; magnesium; potassium; bicarbonate; detergents; DRO; BTEX; isotopic methane and radionuclides; and

- Information related to the anticipated true vertical depth(s) of the formation(s) to be hydraulically fractured and the anticipated pressure range for the proposed HF treatment(s).

---

[18] Such information may be best represented on the maps, cross sections or other graphical representations that must be submitted with the permit application (40 CFR 146.24).

[19] These regulations require the characterization of formation fluids through logging and testing that may be needed given site conditions.

Can Multiple UIC Class II Wells Using Diesel Fuels for HF Be Authorized by One Permit?

Existing requirements in 40 CFR parts 124 and 144 through 147:

An area permit is an option for authorizing injection where there are multiple wells drilled by one owner or operator within a well-defined, localized geologic setting. As provided in 40 CFR 144.33(a), an area permit may be authorized in lieu of an individual permit for each well if the following conditions are met:

- The wells are operated by a single owner or operator;

- The wells are within the same well field, facility site, reservoir, project or similar unit in the same state; and

- The wells are not used to inject hazardous waste.

The regulations at 40 CFR 144.33(b) also specify what must be included in an area permit. Area permits must specify the area within which underground injection is authorized and the requirements for construction, monitoring, reporting, operation and plugging and abandonment for all wells authorized by the permit. As provided in 40 CFR 144.33(c), the area permit may authorize the permittee to construct and operate, convert or plug and abandon additional wells within the permit area provided:

1. The permittee notifies the UIC Program Director at such time as the permit requires;

2. An additional well satisfies the criteria for inclusion in the area permit (40 CFR 144.33(a)) and meets the requirements specified in the permit (40 CFR 144.33(b)); and

3. The cumulative effects of drilling and operation of additional injection wells are taken into account by the UIC Program Director during evaluation of the area permit application and are acceptable to the UIC Program Director.

Recommendations for applying existing requirements to HF activities using diesel fuels:

Below are the EPA's recommendations for applying existing requirements described above regarding issuing area permits for wells where diesel fuels will be used during HF. Recommendations are consistent with the discretion accorded under the existing UIC Class II regulations, and reflect existing UIC requirements for other well classes, voluntary industry standards, state rules, and other model guidelines for hydraulic fracturing.

EPA UIC permit writers should consider issuing area permits for Class II wells using diesel fuels for HF provided that all applicable requirements, including any applicable public notification requirements, are satisfied. Issuing area permits may result in improved permitting efficiency, especially in areas with large numbers of Class II wells using diesel fuels for HF. EPA UIC permit writers should also take into account the total number of proposed wells that will be covered by the area permit when determining the appropriate financial responsibility demonstration to ensure that sufficient resources are available to protect USDWs.

How Should EPA UIC Permit Writers Establish Permit Duration and Apply UIC Class II Requirements After HF at a Well Ceases?

Existing requirements in 40 CFR parts 124 and 144 through 147:

Under the UIC Program, a well may be:

- Permitted as an active injection well for the life of the facility and subject to all applicable Class II requirements (40 CFR 144.36(a));

- Converted out of the UIC Program after injection ceases (meaning the permit duration ends upon conclusion of HF and post-HF monitoring). UIC regulations at 40 CFR 144.36(c) allow a permit to be issued for a duration less than the full allowable term (i.e., the operating life of the facility) indicated at 40 CFR 144.36(a); and 40 CFR 144.51(n) and 144.52(a)(7)(i)(B) allow for conversion of an injection well out of the UIC Program in situations where injection has ceased and production operations are occurring. If a well is converted out of the UIC Program, it is no longer subject to UIC requirements after the permit expires, but may not conduct future permitted underground injection activities (i.e., injection of diesel fuels for HF) unless a new permit is obtained; or

- Managed as a temporarily abandoned (TA) injection well during times when injection ceases or is curtailed. UIC regulations at 40 CFR 144.52(a)(6)(ii) allow for the temporary or intermittent cessation of injection[20] while the permit is active, provided that the owner or operator describes, and the EPA Regional Administrator (RA) approves, actions and procedures that the owner or operator will take to ensure that the well will not endanger USDWs during the period of temporary abandonment.

As described in the section, "Can Multiple UIC Class II Wells Using Diesel Fuels for HF Be Authorized by One Permit," area permits can also be issued per 144.33. For area permits, EPA regulations at 40 CFR 144.51(n) state that the UIC Program Director should be notified before closure of a project, indicating that the duration of the permit should be set so that the area

---

[20] The EPA permit writer has the option of ending the permit after the conclusion of injection or managing the well as TA. Further, regulations state that "temporary or intermittent cessation of injection operations is not abandonment," for the purposes of well closure plans (40 CFR 144.51(o)). Therefore, TA wells remain subject to well closure requirements. For additional guidance, see "Management and Monitoring Requirements for Class II Wells in Temporary Abandoned Status" (US EPA, 1992).

permit does not expire until after the closure of all wells covered by the permit or after the conversion of all wells to oil and gas production (i.e., out of the UIC Program).

Recommendations for applying existing requirements to HF activities using diesel fuels:

The paragraphs below present the EPA's recommended approaches for applying the existing Class II requirements described above regarding setting the permit duration for wells where diesel fuels will be used during HF. Recommendations are consistent with the discretion accorded under the existing UIC Class II regulations, and reflect existing UIC requirements for other well classes, voluntary industry standards, state rules, and other model guidelines for hydraulic fracturing.

EPA UIC Program Directors should consider the following ways of setting the permit duration for an individual well using diesel fuels for HF:

(1) Set a short duration for the permit, as permissible under 40 CFR 144.36(c), that concludes after injection ceases and a non-endangerment demonstration is made. Compliance with UIC permit conditions should be confirmed before the injection permit duration ends and prior to releasing it from UIC requirements. Note that, as stated above, under this recommendation, an owner or operator of a production well wishing to refracture the formation using diesel fuels after the conclusion of the UIC permit would need to receive a new, approved UIC permit before refracturing can occur. The EPA recommends that the duration of a permit that is less than the full allowable term still allow adequate time to collect monitoring data that demonstrates that injection during the HF operation has not endangered USDWs in the project area. This timeframe is likely to vary, depending on site-specific factors.

(2) Manage the well as temporarily abandoned during periods of oil or gas production (e.g., when no injection is occurring). This option may be preferable in situations where the well owner or operator plans to refracture the formation using diesel fuels at some point in the future. During a period of temporary cessation of injection, the UIC Program Director may authorize alternative or reduced requirements for mechanical integrity, operation, monitoring and reporting other than those required in 40 CFR 146 and 144.52, making them more appropriate to the short-term nature of HF, to the extent that changes in requirements will not result in an increased risk of movement of fluid into a USDW (40 CFR 144.16). A well may be considered as meeting the conditions of 40 CFR 144.16 if:

- It is not injecting into, or through or above a USDW, or

- It is injecting into, or through or above a USDW, but has a ZEI[21] that is smaller than the radius of the well when computed using the formula at 40 CFR 146.6(a).

Either situation could occur when the well is producing (e.g., when no injection is occurring) and the injection rate is zero. When managing a well as TA, the EPA UIC permit writer should use

---

[21] The ZEI is the lateral area in which the pressures in the injection zone may cause injection or formation fluid to migrate into a USDW (further described in Appendix B).

his or her authorized discretion under 40 CFR 144.52(a)(9) to tailor permit conditions on a case-by-case basis. Permit writers may consider making changes in a number of areas including: frequency of mechanical integrity testing, monitoring for ground water quality, injection pressure, flow rate and cumulative volumes monitoring and certain reporting requirements. However, permit conditions should still ensure that a mechanical integrity test (MIT) is conducted just prior to returning the well to active injection. In conjunction with the MIT test, pressure tests and cement bond logs should be submitted to the Director prior to refracturing the well using diesel fuels.

For area permits, EPA UIC Program Directors should ensure that wells are in compliance with all aspects of the UIC area permit prior to releasing any from UIC Program requirements. The EPA UIC Program Director should review the area permit conditions after the first few wells are drilled and hydraulically fractured to make adjustments, as needed, based upon any new data collected. Thereafter, permit conditions should be reviewed at least once every five years for the duration of the area permit.

Properly closing an injection well is critical to assuring the long-term prevention of contamination of USDWs by eliminating a potential pathway, or pathways, for contamination. Both the UIC Program and state oil and gas programs require well closure. Coordination should be feasible because state oil and gas programs typically require closure, plugging and abandonment activities for production wells that are similar to what the UIC Program requires for underground injection wells. The owner or operator of a production well who wishes to refracture a well using diesel fuels that had been released from the UIC Program by being fully converted to production would need to submit a new UIC permit application.

How Do the Area of Review (AoR) Requirements at 40 CFR 146.6 Apply to Wells Using Diesel Fuels for HF?

Existing requirements in 40 CFR parts 124 and 144 through 147:

The AoR is the area surrounding an injection well and is defined at 40 CFR 146.3. The AoR must be determined by one of two methods according to the criteria set forth in 40 CFR 146.6: (1) determining the ZEI, or (2) using a minimum one-quarter (¼) mile fixed radius around the well. In the case of an area permit, the AoR is the project area plus a circumscribing area the width of which is either ¼ of a mile or a number that is calculated (i.e., ZEI). The EPA UIC permit writer may solicit input as to which method is most appropriate for each geographic area or field. If the AoR is determined by modeling, the applicable radius is the result of the modeling, even if it is less than one-quarter (¼) mile.

Delineating and evaluating an AoR helps to ensure that there are no conduits in the vicinity of the injection well that could enable fluids to migrate into USDWs and identifies conduits which must be appropriately addressed by corrective action. Before proceeding with the project, the owner or operator must define the appropriate AoR, assess that area for conduits of potential fluid movement and if necessary, perform corrective action, such as the plugging of improperly abandoned and orphaned wells, or re-siting of the planned well to account for any conduits that could potentially cause migration of contaminants into USDWs.

Recommendations for applying existing requirements to HF activities using diesel fuels:

The paragraphs below present the EPA's recommended approaches for applying the existing Class II requirements described above, for defining the AoR for wells where diesel fuels will be used during HF. Recommendations are consistent with the discretion accorded under the existing UIC Class II regulations, and reflect existing UIC requirements for other well classes, voluntary industry standards, state rules, and other model guidelines for hydraulic fracturing.

EPA UIC Program Directors should modify the one-quarter (¼) mile fixed radius approach to delineating the AoR so that it prevents endangerment of USDWs. Site-specific AoR determinations are needed to address the full extent, shape and size of the AoR for HF projects using diesel fuels based on consideration of geology, operations and directional drilling, which typically extends beyond one-quarter mile from the wellhead.

Modifying the fixed radius approach may require the EPA UIC permit writer to review past HF activities in each geographic area or field and consult with the owner or operator about the design and anticipated results for the fracturing operation. Information needed in determining the appropriate AoR delineation method includes three-dimensional well orientation and anticipated fracture length. In addition, multiple wells co-located on the same well pad introduce complexities into the AoR delineation and assessment process. Thus, owners or operators using multi-well pads should include length and angle of each directional completion, fracture length and an estimation of how closely the fractured zone approximates a porous medium. Appendix B presents methods for calculating the AoR for individual directionally completed wells and multiple directionally completed wells and provides further discussion of the limitations of the Theis equation in settings where the well is directionally completed.

The EPA recommends against using the modified Theis equation found at 40 CFR 146.6 to determine the ZEI for directional wells because directional wells do not meet the equation's assumptions for the well, the aquifer conditions and the similarity of hydraulic properties between the injectate and the in situ groundwater. However, computational models may be a desirable option. A further brief discussion of ZEI modeling is found in Appendix B: Methods for Calculating the Area of Review. Appendix B provides clarifications of 40 CFR 146.6 for the purpose of delineating an AoR for a directionally completed well.

How Do the Class II Well Construction Requirements Apply to HF Wells Using Diesel Fuels?

Existing requirements in 40 CFR parts 124 and 144 through 147:

Specific construction requirements for Class II injection wells, including Class II HF wells using diesel fuels, are found at 40 CFR 146.22. These requirements establish that Class II wells must be cased and cemented in a manner that prevents the movement of fluids into or between USDWs for the life expectancy of the well. EPA UIC permit writers must consider the following factors in determining casing and cementing requirements for new Class II HF wells using diesel fuels:

- Geology of the injection and confining zones including the estimated formation fracture pressure;

- Depth from surface to the injection zone and to the bottom of each USDW down to and including the lowermost USDW; and

- Proposed operating procedures including maximum and average injection pressures (40 CFR 146.22(b)(1)(iii)).

Recommendations for applying existing requirements to HF activities using diesel fuels:

EPA UIC Program Directors should consider the following recommendations for applying the existing Class II requirements described above to ensure that the well is designed and constructed for the unique geologic environment and planned use of diesel fuels for HF operations. Recommendations are consistent with the discretion accorded under the existing UIC Class II regulations, and reflect existing UIC requirements for other well classes, voluntary industry standards, state rules, and other model guidelines for hydraulic fracturing.

EPA UIC Program Directors should ensure that a combination of casing and cement isolates the lowermost USDW encountered in the borehole from HF target formation(s) when specifying casing and cementing requirements for Class II wells using diesel fuels for HF (40 CFR 144.52(a)(9)). Isolating the lowermost USDW encountered through the use of casing and cement along the entire borehole is consistent with federal requirements for several classes of injection wells, is recommended in API guidance[22] and is a requirement for HF wells in several states. To ensure that the well cement has been emplaced properly and zonal isolation has been achieved, appropriate logs and other test results such as sonic, temperature, cement bond or other cement evaluation logs (CELs) and fracture finder logs should be considered during the drilling and construction of Class II HF wells using diesel fuels (40 CFR 146.22).

EPA UIC Program Directors should consider the use and placement of centralizers when specifying the cementing requirements for Class II wells using diesel fuels for HF (40 CFR 144.52(a)(9)). Centralizing the casing in the borehole helps to ensure that the casing is more uniformly encased by cement during the cementing operation which, in turn, helps ensure zonal isolation that protects USDWs from fluid migration along the wellbore.

To ensure that appropriate precautions are taken to address the high injection pressures needed for HF, EPA UIC Program Directors should consider requesting the following information to assist in specifying casing and cementing requirements:

- A description of the geologic formations overlying the production zone and whether they might contain gas, oil or other potentially mobile contaminants that should be isolated

---

[22] *API Guidance* recommends that surface casing, at a minimum, be set at least 100 feet below the deepest USDW encountered (API, 2009).

from the well by cement. Isolating zones of potential contaminants would decrease the risk of endangerment to USDWs from movement of contaminants into nearby USDWs;

- A review of well construction plans to consider and address potential pathways for fluid migration between any gas-bearing zones and USDWs including identification of layers that may release hydrocarbons into the drilling fluids and into USDWs. For example, if surface casing is not installed properly prior to drilling, shallow gas may migrate upwards through the borehole and may potentially impact USDWs;

- The physical and chemical characteristics of the formation fluids in the injection zone and the proposed characteristics of the well such as the size of the borehole, which are needed to determine appropriate construction materials for the use and life of the well. Construction materials should maintain integrity over the life of the well in order to protect USDWs;

- Location and operating procedures of other active injection wells or wells undergoing HF in the AoR or nearby injection zones. Pressures external to the well coupled with injection pressure may cumulatively affect the integrity of the construction materials and fracture pressure of the injection zone. Exceeding the capability of the construction materials would cause failure of mechanical integrity and possible leaks of fluids into USDWs. Exceeding the fracture pressure of the injection zone risks fracturing confining zones and creating conduits for fluids to move into USDWs;

- Data on sizes and grades of the casing string and classes of cement to be used in construction (40 CFR 146.22(b)-146.22(g));[23]

- The proposed cementing plan to ensure proper cement design and volume. Related information of particular importance includes the capability of the typically lower-density "lead" cement to adequately isolate overlying USDWs, which would assist in evaluating if the higher-density and compressive-strength "tail" cement coverage should be modified (placed higher) to effectively isolate and afford appropriate protection of overlying USDWs; and

- Additional information to ensure that long, multi-well pad horizontal wells will be constructed in a protective manner.

The EPA UIC permit writer may also consider additional testing requirements to demonstrate that the well maintains mechanical integrity before, during and after the use of diesel fuels for HF injection event (40 CFR 144.52), as described in the section titled "How Do the Class II Well Operation, Mechanical Integrity, Monitoring, and Reporting Requirements Apply to HF Wells Using Diesel Fuels?"

---

[23]API recommends that casing used in oil and gas wells that will be hydraulically fractured meet API standards, including *API Specification 5CT* (API, 2005).

Different considerations may apply for already constructed wells. (See "How Do the Class II Well Construction Requirements Apply to Already Constructed Wells Using Diesel Fuels for HF?" for applicable information on already constructed wells.)

How Do the Class II Well Construction Requirements Apply to Already Constructed Wells Using Diesel Fuels for HF?

Existing requirements in 40 CFR parts 124 and 144 through 147:

Wells constructed prior to issuance of this guidance (i.e., already constructed wells) may have been constructed and operated under requirements other than the federal UIC Class II requirements. EPA UIC permit writers, under 40 CFR 146.22(c), may authorize an already constructed well for Class II injection activities if the owner or operator can demonstrate that injection will not result in movement of fluids into a USDW so as to create a significant risk to the health of persons. The demonstration might include requiring the owner or operator to obtain downhole logs and internal and external MITs prior to any HF injection activities using diesel fuels to ensure that well construction will prevent fluid migration into USDWs.

Recommendations for applying existing requirements to HF activities using diesel fuels:

The EPA UIC permit writer should consider the following recommendations for applying the existing requirements described above, when permitting already constructed wells as UIC wells for HF using diesel fuels. Recommendations are consistent with the discretion accorded under the existing UIC Class II regulations, and reflect existing UIC requirements for other well classes, voluntary industry standards, state rules, and other model guidelines for hydraulic fracturing.

EPA UIC Program Directors should ensure the owner or operator applies relevant construction-related requirements to already constructed Class II HF wells using diesel fuels to protect USDWs during injection for HF using diesel fuels (40 CFR 144.52(a)(9)). EPA UIC permit writers should consider consulting with the oil and gas agency that may have permitted the well (e.g., during past production operations) to learn about the well's compliance history or other relevant information in order to make permit determinations about the appropriateness of permitting the well for UIC Class II diesel fuels HF use.

Some already constructed oil and gas wells may not provide an adequate level of protection for USDWs when undergoing the use of diesel fuels for HF-related injection due to either the age of the well or to less stringent well construction standards that were in place when the well was constructed. For example, an older well may not be cemented to the lowermost USDW encountered or construction may not be adequate to withstand proposed injection pressures anticipated during the use of diesel fuels for HF. If a well does not provide adequate protection for USDWs, then the EPA UIC permit writer should require the owner or operator to perform actions to ensure that USDWs are not endangered. Actions to repair a well include, but are not limited to, replacing the injection well tubing or cementing across specific sections of the well that intersect potentially vulnerable formations to decrease the risk of fluid movement. If corrective measures are not sufficient to protect USDWs, the EPA UIC permit writer should not issue a permit, consistent with 40 CFR 144.12.

How Do the Class II Well Operation, Mechanical Integrity, Monitoring and Reporting Requirements Apply to HF Wells Using Diesel Fuels?

Well Operation

Existing requirements in 40 CFR parts 124 and 144 through 147:

Injection well operating requirements for Class II wells are found at 40 CFR 146.23(a). They require that, at a minimum, injection pressure should be limited so that injection does not cause the propagation of new fractures in confining zone(s) adjacent to USDWs. The purpose of these requirements is to ensure that the integrity of confining zones protecting USDWs is maintained and that injection pressures do not cause the movement of injection or formation fluids into USDWs. In addition, the EPA UIC permit writer should consider the following recommendations when permitting these wells as UIC wells for HF using diesel fuels.

Recommendations for applying existing requirements to HF activities using diesel fuels:

The paragraphs below present the EPA's recommended approaches for applying the existing requirements described above, regarding the operation of wells where diesel fuels will be used during HF. Recommendations are consistent with the discretion accorded under the existing UIC Class II regulations, and reflect existing UIC requirements for other well classes, voluntary industry standards, state rules, and other model guidelines for hydraulic fracturing.

EPA UIC Program Directors should consult with the owner or operator about the design and anticipated results of a proposed fracturing operation. It is important to establish operating requirements that are appropriate for the proposed use of diesel fuels for HF operations and that account for past HF activities in each geographic area or field. Historical production and HF activities may have created fracture networks that will interact with future HF operations using diesel fuels. Awareness of the existing fracture network and anticipation of fracture interactions when designing new HF operations will decrease the risk of endangerment to USDWs. The consultation increases the ability for owners or operators to incorporate recommended approaches into the modeling often used to design and determine parameters of a proposed use of diesel fuels for HF operation.

EPA UIC Program Directors should consider construction design and geologic conditions when determining the maximum injection pressure for a UIC permit ((40 CFR 144.52(a)(9) and 40 CFR 146.23(a)(1)). EPA UIC permit writers should examine the fracture gradient of the injection zone and other intervening geologic zones to determine fracture pressure and to avoid damage to the confining zone, which acts as a barrier to protect USDWs. Calculations of maximum injection pressure should also consider the properties of the construction materials to withstand HF.

EPA UIC Program Directors should ensure that wells used for diesel fuels for HF incorporate appropriate controls (e.g., pressure limitations) so that integrity of the confining zone(s) protecting USDWs are maintained in order to comply with 40 CFR 146.23. Many oil and gas extraction practices tend to reduce pressures in the formation, and typical oil and gas production regulations are designed for these circumstances while typical injection activities, including the use of diesel fuels for HF, in general, increase formation pressures. UIC Program regulations and associated permit conditions generally address risks associated with pressure increases.

Mechanical Integrity Testing

MITs ensure that the protective physical components of the well are competent prior to injection and over the life of the injection well. High injection pressures, such as those occurring during HF, have the potential to damage the mechanical integrity of the well causing leaks, which may allow for the migration of fluids into USDWs. Injection well integrity must be maintained at all times during HF using diesel fuels and during any subsequent refracturing events.

Existing requirements in 40 CFR parts 124 and 144 through 147:

The mechanical integrity requirements, found at 40 CFR 146.8, describe methods for demonstrating mechanical integrity of well tubular components, or internal mechanical integrity, and the cement around the well casing, or external mechanical integrity, over the life of the injection well.

Recommendations for applying existing requirements to HF activities using diesel fuels:

The paragraphs below present the EPA's recommended approaches for applying the existing Class II requirements described above with regard to ensuring mechanical integrity of wells where diesel fuels will be used during HF. Recommendations are consistent with the discretion accorded under the existing UIC Class II regulations, and reflect existing UIC requirements for other well classes, voluntary industry standards, state rules, and other model guidelines for hydraulic fracturing.

To account for the unique nature of diesel fuels HF injection—including high pressures involved, high volumes of fluids, and the use of diesel fuels—EPA UIC permit writers should also consider incorporating into permit conditions the procedures listed below consistent with 40 CFR 144.52(a)(9) and 40 CFR 146.8(a)(1) to ensure that there is no significant leak in the casing and when applicable, tubing and packer through the following methods:

- Perform casing integrity tests of casing strings (surface, intermediate and when necessary, the production casing) prior to drilling out beneath each casing shoe at a pressure that will determine if the casing integrity is adequate to meet well design and construction objectives. Report tests results to the EPA UIC Program Director along with the well completion report (40 CFR 144.51(m));

- Perform formation pressure tests immediately after drilling out the surface, intermediate, and production casings and report tests results to the EPA UIC Program Director along with the well completion report (40 CFR 144.51(m));

- Conduct a casing integrity test of the production casing, or a Standard Annular Pressure Test (SAPT) for wells with a tubing and packer arrangement, at pressures equal to or exceeding the maximum expected pressure during HF operations[24] prior to perforating and fracturing the well to ensure that the pressure during stimulation does not compromise the integrity of the casing. The EPA UIC Program Director should consider the production casing integrity test/SAPT prior to approving HF operations using diesel fuels (40 CFR 146.24(c)); and

- Equip the wellhead with pressure recording devices on all available annuli and injection strings with a gauge pressure rating adequate to monitor well construction performance during HF operations that use diesel fuels.

To account for the unique nature of diesel fuels HF injection, EPA UIC permit writers should also consider incorporating into permit conditions the procedures listed below consistent with 40 CFR 144.52(a)(9) and 40 CFR 146.8(a)(2) to ensure that there is no significant fluid movement in channels adjacent to the well bore through cement integrity evaluation using the following methods:

- During well construction, monitor and record the volume, flow rate, density and treating pressure of cement operations; and

- Submit a CEL with the notice of completion of construction (40 CFR 144.51 (m)) for review and approval by the EPA UIC Program Director; CELs can provide an assessment of the presence or absence of cement and how effectively cement is bonded to the pipe. Acceptable CELs include, but are not limited to: radial cement bond log, ultrasound imager, magnetic resonance imager and isolation scanner.

---

[24] If testing at the maximum expected fracturing pressure is reasonably expected to harm the production formation below the casing shoe, the UIC Program Director may authorize testing at a lower pressure.

To account for the unique nature of diesel fuels HF injection, EPA UIC permit writers should also consider the following consistent with 40 CFR 144.52(a)(9) and 40 CFR 146.8 to assess mechanical integrity and ensure USDW protection during diesel fuels HF:

- Request the permittee to report to the EPA, verbally within 24 hours and with written confirmation that includes certification and documentation of remedial cementing[25] within 48 hours if a casing integrity test, a formation pressure test, cementing records or a CEL provides indication of inadequate cementing or a failure. Certification and documentation of remedial cementing that indicates adequate cement bonding must be submitted to the EPA for review and approval prior to resumption of operations;

- Request additional mechanical integrity testing such as noise logs, oxygen activation logs, temperature logs and other logs approved by the EPA UIC Program Director if the results of a diesel fuels HF well's pressure testing and/or CELs do not confirm that there is no significant leak in the casing, tubing or packer of the injection and that there is no significant fluid movement into a USDW through vertical channels adjacent to the injection well bore (40 CFR 146.8);

- Once a diesel fuels HF well has been converted to production, waive the Class II requirement at 40 CFR 146.23(b)(3) to conduct MITs every five years during the life of the well, if doing so will not result in an increased risk of movement of fluids into USDWs per discretion at 144.16; and

- As necessary, adjust requirements for mechanical integrity testing to confirm compliance with UIC permit conditions and non-endangerment of USDWs before expiration of the injection permit (40 CFR 144.51 (q)(1)).

Monitoring and Reporting

The collection and review of monitoring data enables EPA UIC permit writers to confirm that the well is operating safely as expected and within the established parameters of the permit.

Existing requirements in 40 CFR parts 124 and 144 through 147:

Existing Class II regulations for monitoring and reporting before, during and after a Class II well commences operation, are found at 40 CFR 146.23(b) and 40 CFR 144.51 and are summarized in Table 1. The UIC permit writers can use their discretion at 40 CFR 144.52(a)(9) to allow flexibility in setting permit conditions for monitoring and reporting in certain site-specific conditions where alternative approaches can be demonstrated to be as effective at preventing migration of fluids into USDWs. Also, the UIC Program Director under (40 CFR 144.16) may

---

[25] Remedial cementing operations may be done in accordance with methods pre-approved by the EPA UIC Program Director as provided under UIC Class II permit conditions. Per the EPA UIC Director's discretion, certification and documentation of pre-approved remedial cementing operations may be included in the well completion report (40 CFR 144.51 (m)). Remedial cementing operations not included in the pre-approved methods shall be submitted for approval to the EPA UIC Program Director before proceeding.

authorize less frequent monitoring during certain phases of the permit such as during production periods for diesel fuels HF wells under temporary abandonment status.

Recommendations for applying existing requirements to HF activities using diesel fuels:

Below are the EPA's recommended approaches for applying the existing Class II requirements described above, with regard to monitoring and reporting for wells where diesel fuels will be used during HF. Recommendations are consistent with the discretion accorded under the existing UIC Class II regulations, and reflect existing UIC requirements for other well classes, voluntary industry standards, state rules, and other model guidelines for hydraulic fracturing.

EPA UIC Program Directors should modify monitoring and reporting protocols, consistent with their authorized discretion under 40 CFR 144.52(a)(9), so that the permit writer has adequate information to determine that each planned HF operation using diesel fuels will not endanger USDWs, including:

- Monitoring pump rate, pressure, volume and viscosity of the fracturing fluid to evaluate the results of the diesel fuels HF operation, such as fracture vertical length and lateral extent to confirm the protection of USDWs during diesel fuels HF. Data that can be collected during the treatment operation to monitor and control operations in real-time include continuously monitored surface injection pressure, injection rate and volume, slurry rate and percentage proppant. An owner or operator may also choose to use microseismic and tiltmeter surveys as suggested in API guidance[26] to achieve real-time mapping of a HF treatment in progress;

- Allowing flexibility in monitoring and reporting protocols to address the intermittent, or infrequent, nature of HF using diesel fuels wells while remaining protective of USDWs; and

- Utilizing alternative and supplemental monitoring data (e.g., micro-seismic or tiltmeter data), where appropriate.

---

[26] API Guidance (API, 2009).

Table 1. Existing Class II Diesel Fuels Hydraulic Fracturing Well Monitoring and Reporting Requirements[27]

	Required Activity	Required Timing	Recommended Timing	Purpose
Operations and Monitoring	Conduct appropriate logging and testing to assess USDWs, injection zones, confining zones and adjacent formations; prepare a report synthesizing logging and testing results [40 CFR 146.22(f)]	During drilling and construction	Same	Provides data and information on the subsurface, including the location of injection zones, confining zones and adjacent formations; informs permitting decisions to prevent migration of injected fluids into USDWs and ensure USDW protection
	Monitor the nature of injected fluids [40 CFR 146.23(b)(1)]	At a frequency sufficient to yield data representative of the fluid characteristics	Same	Provides an understanding of the potential risks of fluid migration
	Monitor injection pressure, flow rate and cumulative volume [40 CFR 146.23(b)(2) (ii)]	At least monthly	Continuously during diesel fuels HF injection; the EPA UIC permit writer may use discretion to adjust the frequency of monitoring thereafter	Ensures protective injection well operational parameters are met
MIT	Conduct mechanical integrity testing [40 CFR 146.8(b)(1); 40 CFR 146.23(b)(3)]	Prior to being authorized to inject and at least once every five years during the life of a project	Prior to being authorized to inject and if pressure testing and/or CELs cannot confirm absence of significant leak in casing, etc., and absence of fluid movement into USDWs	Determines well component integrity and/or if corrective action is needed to prevent vertical migration through the well bore
Pre-permit Expiration	Conduct pre-permit expiration monitoring [40 CFR 144.52(a)(9)]	At location and frequency as approved by the EPA UIC Director in the pre-permit expiration monitoring plan and in permit conditions	Same	Establishes groundwater quality conditions before and after diesel fuels HF to demonstrate non-migration of fracturing fluids and detect potential changes in quality resultant from the fracturing activity

---

[27] This table lists current requirements for monitoring and reporting and adjustments for diesel fuels HF Class II wells that the EPA recommends that permit writers consider.

	Required Activity	Required Timing	Recommended Timing	Purpose
Reporting and Record-Keeping	Report any emergency or noncompliance event which may endanger human health or the environment [40 CFR 144.51(k)(6)]	Verbally, within 24 hours; in writing, within five days of an emergency or noncompliance event	Same	Provides for timely initiation of remedial action
	Notify the EPA UIC Program Director that construction is complete and await approval before commencing injection [40 CFR 144.51(m)]	After well construction completion	Same	Provides the EPA UIC Program Director information to ensure well construction is protective of USDWs prior to operation
	Report information collected under 146.23(b)(1) before, during and after a Class II well (including Class II HF wells using diesel fuels) commences operation [40 CFR 146.23(b)]	Varies, depending on type and characteristics of the activity being monitored	Same	Ensures maintenance of well integrity so that injected fluids do not migrate into USDWs; informs remedial action, if needed
	Submit a summary report of all monitoring[28] [40 CFR 146.23(c)(1) & (2)]	Annually	As determined by permit conditions and waived at the discretion of the EPA UIC Program Director thereafter	Allows the EPA UIC Program Director to review activities and ensure the permit conditions are met
Record Retention	Retain all calibration and maintenance records; original strip chart recordings for continuous monitoring; copies of all reports required by the permit and data used to complete the permit application; and monitoring records on the nature and composition of all injected fluids [40 CFR 144.51(j)(2)(i) & 40 CFR 144.51(j)(2)(ii)]	Retain for three years from the date of the sample, procedure, measurement, report[29] or application. Retain information on the nature and composition of all injected fluids until three years after the completion of any plugging and abandonment,	Same	Confirms safe and protective injection; informs future activities in the AoR and any necessary remedial action

---

[28] Owners or operators of enhanced recovery wells may report on a field or project basis rather than an individual well basis.

[29] For EPA-administered programs, the owner or operator shall retain records beyond three years, unless records are delivered to the RA or the RA gives written approval to discard them.

	Required Activity	Required Timing	Recommended Timing	Purpose
	Maintain results of all monitoring [40 CFR 146.23(b)(4)]	Until the next permit review	Same	Confirms USDW protection during injection; informs future activities in the AoR and any necessary remedial action

How Do the Class II Financial Responsibility Requirements Apply to Wells Using Diesel Fuels for HF?

Existing requirements in 40 CFR parts 124 and 144 through 147:

Like other classes of injection wells, the Class II regulations require a demonstration of financial responsibility (or available resources) before any operation can be performed (including the use of diesel fuels for HF operations). Regulations for Class II wells require a demonstration of financial responsibility to cover the costs of closing, plugging and abandoning an underground injection well (40 CFR 144.52(a)(7)). The demonstration and maintenance of financial responsibility is a permit condition that is required until: (a) the well is closed in accordance with an approved plugging and abandonment plan; (b) the well has been converted to production (i.e., no longer injecting for the purposes of the UIC Program); or (c) the transferor of a permit has received notice from the EPA UIC Program Director that the new permittee has demonstrated financial responsibility for the well (40 CFR 144.52(a)(7)). Submission of surety bonds, financial statements or acceptable materials to show evidence of financial responsibility is required.

EPA UIC permit writers may periodically require revisions to the financial responsibility demonstration. This includes an update to the cost estimate of the resources needed to plug and abandon the well to reflect inflation of such costs.

Class II wells using diesel fuels for HF operations will at some point cease injection and begin oil and gas production. Financial responsibility must be maintained under the UIC permit until the well has been properly closed and plugged or for the duration of the permit in cases where wells are converted out of the UIC Program and into oil and gas production. (See "How Should EPA UIC Permit Writers Establish Permit Duration and Apply UIC Class II Requirements After HF at a Well Ceases?" for applicable information on permit duration and well conversion.)

Recommendations for applying existing requirements to HF activities using diesel fuels:

The paragraphs below present the EPA's recommended approaches for applying the existing requirements described above, to ensuring financial responsibility for wells where diesel fuels will be used during HF. Recommendations are consistent with the discretion accorded under the existing UIC Class II regulations, and reflect existing UIC requirements for other well classes, voluntary industry standards, state rules, and other model guidelines for hydraulic fracturing.

The EPA UIC Program Director should thoroughly examine proposals that use a financial test or corporate guarantee for self-insurance. Compared to third-party instruments (e.g., trust fund, surety bond, letter of credit), self-insurance may pose a higher risk of instrument failure

(US EPA, 2005; U.S. Government Accountability Office, 2005). If an owner or operator selects self-insurance, EPA UIC permit writers should evaluate whether the risk of instrument failure is acceptable for ensuring that USDWs will not be endangered.

The EPA UIC Program Director should include coverage for the total number of wells in an area permit for Class II wells using diesel fuels for HF (i.e., the sum of costs for each well covered by an area permit) when determining the extent of financial responsibility required. An acceptable financial responsibility demonstration will indicate that the face value of the financial instrument (i.e., third party financial instruments or self-insurance demonstration) meets or exceeds the plugging costs specified in the Plugging and Abandonment Plan (EPA Form 7520-14) for all wells.

The EPA UIC Program Director should ensure that owners or operators refer to previously published guidance on the EPA-administered UIC Programs for additional context on the recommendations related to financial responsibility with respect to the use of diesel fuels for HF described in this guidance (US EPA, 1990).

What Public Notification Requirements or Special Environmental Justice (EJ) Considerations are Recommended for Authorization of Wells Using Diesel Fuels for HF?

Requirements in 40 CFR parts 124 and 144 through 147:

Public notification requirements for all UIC well classes are addressed in 40 CFR Part 124. Under these requirements, the EPA UIC Program Director must give notice to the public of all permit actions (including those for HF activities using diesel fuels), including when a permit has been tentatively denied, a draft permit has been prepared, a hearing has been scheduled or an appeal has been granted. The public must be given 30 days to comment on a draft permit and 30 days' notice of a planned hearing (40 CFR 124.10). During the 30-day comment period for a draft permit, any interested person may request a hearing (40 CFR 124.11). Public notice of a public hearing may be given at the same time as public notice of the draft permit, and the two notices may be combined (40 CFR 124.10(b)). The public notification requirements were established to enable interested stakeholders to give input into the UIC permitting process.

Recommendations for applying existing requirements to HF activities using diesel fuels:

Below are the EPA's recommendations for applying the existing requirements described above, with regard to improving public information available about the use of diesel fuels for HF operations and incorporating environmental justice (EJ) concerns. Recommendations are consistent with the discretion accorded under the existing UIC Class II regulations, and reflect existing UIC requirements for other well classes, voluntary industry standards, state rules, and other model guidelines for hydraulic fracturing.

The owner or operator and EPA UIC Program Director should begin planning for public notification as soon as a new injection well is proposed to give the maximum amount of time for effective communication while not affecting the project schedule. Public participation will help permitting authorities understand public concerns about these projects. Public participation activities will also give the public an opportunity to gain a clearer understanding of the benefits and risks of the planned use of diesel fuels for HF activity. By beginning outreach early, both the EPA UIC permit writer and the owner or operator will have more flexibility to consider and address stakeholder concerns. Earlier stakeholder outreach can help mitigate controversial issues and avoid litigation and project delays. One way to achieve earlier public notification is to build on requirements at 40 CFR 144.31(e)(9), which specify that permit applicants to the EPA-administered programs should identify and submit with the permit application the names and addresses of all land owners within one-quarter mile of the facility boundary, unless waived by the EPA UIC Program Director. The EPA UIC permit writer could request owners or operators to obtain land owner contact information required in the permit application and also send out project details to local land owners and nearby public officials, including public water supply system operators, regarding the proposed use of diesel fuels for a HF project in advance of submitting the permit application.

Other options EPA permit writers could consider, include, but are not limited to:

- Scheduling a hearing concurrently with the public notice of draft permit in areas where hearing requests are expected; and

- Coordinating application submission for multiple permits from multiple owners or operators to issue one public notice and hold one comment period and/or hearing for multiple permits in a given production area or similar geographic delineation.

The EPA UIC Program Director should make available on the EPA website the draft permit as specified by 40 CFR 124.6 including the contact information for an EPA official to whom members of the public could direct their comments. If the UIC Program Director tentatively decides to issue a UIC permit, a draft permit must be prepared and publicly noticed. The EPA has historically made these draft permits available through a variety of methods. Draft permits contain information that the public is often most concerned about: permit conditions, monitoring and reporting requirements, compliance schedule, corrective actions and more. In addition, all draft permits are required to be accompanied by a statement of basis (40 CFR 124.7) or fact sheet (40 CFR 124.8).

The EPA UIC Program Director and owners or operators should make a special effort to consider Environmental Justice in the permitting process for the use of diesel fuels for HF. The following sub-section, "Incorporating Environmental Justice Considerations," provides a description of how this could be done.

Incorporating Environmental Justice Considerations

Presidential Executive Order 12898, *Federal Actions to Address Environmental Justice in Minority Populations and Low-Income Populations* (59 FR 7269, Feb. 16, 1994), states that "federal agencies shall make achieving environmental justice part of its mission by identifying and addressing, as appropriate, disproportionately high and adverse human health or environmental effects of its programs, policies and activities on minority populations and low-income populations in the United States and its territories…."

The EPA's comprehensive *Plan EJ 2014: Considering Environmental Justice in Permitting* is the agency's roadmap to integrating EJ into its programs and policies. Plan EJ 2014[30] is intended to enable environmental justice (EJ) communities to have full and meaningful access to the permitting process and to develop permits that address EJ issues to the greatest extent practicable. This is the implementation plan for developing a suite of cohesive tools and providing a public database of many other tools to serve as a resource for the EPA and all interested stakeholders to utilize during the permitting process. Potential tools in development include guidance, best practices and fact sheets on permit processes, public involvement and communication, permit conditions and interagency protocols. EPA UIC permit writers should consult Plan EJ 2014 and other resources and work with owners or operators to reduce or mitigate any potential EJ impacts of a proposed use of diesel fuels for HF activity. Up-to-date information on completed and pending EJ tools and details on the EPA's progress on implementing EJ 2014 are available in the Plan EJ 2014: Progress Reports[31]. Appropriate efforts in this regard are particularly important in light of the widespread interest in, or concern about impacts of HF on communities.

Implementation

EPA Regional Offices directly implementing the UIC Class II Program should consider the recommendations in this guidance in permitting HF activities that use diesel fuels to ensure protection of USDWs. However, EPA permit writers have the discretion to consider alternative approaches that are consistent with the statutory and regulatory requirements. EPA Regional Offices should continue to coordinate with state oil and gas programs and the appropriate BLM office and to establish a mechanism to inform owners or operators of applicable UIC Program requirements and application deadlines. In addition, EPA Regional Offices should collaborate with appropriate regulatory entities to streamline permitting (e.g., between the EPA and BLM on BLM-managed lands or with state agencies) such as sharing data where requirements and reporting timeframes are compatible for coordination among the various permitting authorities.

---

[30] US EPA, http://www.epa.gov/environmentaljustice/plan-ej/permitting.html
[31] US EPA, http://www.epa.gov/compliance/ej/resources/policy/plan-ej-2014/plan-ej-progress-report-2013.pdf

---

References

API. 2005. API Specification 5CT – Specification for Casing and Tubing. 8th Edition.

API. 2009. Hydraulic Fracturing Operations – Well Construction and Integrity Guidelines. American Petroleum Institute Guidance Document HF1. 1st Edition. http://www.api.org/policy/exploration/hydraulicfracturing/index.cfm#primer

API. 2012. High Production Volume (HPV) Chemical Challenge Program: Gas Oils Category Analysis Document and Hazard Characterization. http://www.petroleumhpv.org/docs/gas_oil/2012_nov15_Gas%20Oils%20CAD%20Final%20Standard%2010_24_2012.pdf

ACS. 2011. CAS Registry. American Chemical Society. http://www.cas.org/expertise/cascontent/registry/regsys.html.

DeVine C.S., W.D. Wood, M. Sherkarchian, and B.R. Hunnicutt. 2003. New Environmentally Friendly Oil-Based Stimulation Fluids. Society of Petroleum Engineers Paper 84576, SPE Annual Technical Conference, Denver, CO, 5-8 October 2003.

Geraghty and Miller, Inc. 1980. Mechanical Integrity Testing of Injection Wells. April 30, 1980. http://nepis.epa.gov/Exe/ZyPURL.cgi?Dockey=9100EYTR.txt

McCabe, M.A., J.M. Terracina, and R.A. Kunzi. 1990. Continuously Gelled Diesel Systems for Fracturing Applications. CIM/SPE International Technical Meeting.

Rae, P., and G. DiLullo. 1996. Fracturing Fluids and Breaker Systems – A Review of State of the Art. SPE Eastern Regional Meeting.

Safe Drinking Water Act, Underground Injection Control Regulations, 45 Fed. Reg. 42472 (June 24, 1980).

Shibley, J.A., and R.A.F Leonard. 1987. Improved Well Productivity Realized by Fracturing with Frac Oil as Compared to Diesel. Journal of Canadian Petroleum Technology. Vol. 26 No. 1.

US Department of Energy Secretary of Energy Advisory Board (US DOE). 2011. Shale Gas Production Subcommittee 90-Day Report. http://www.shalegas.energy.gov/resources/081811_90_day_report_final.pdf.

US EPA. 1977. An Introduction to the Technology of Subsurface Wastewater Injection. Chapter 7 ("Subsurface Wastewater Injection"). nepis.epa.gov/Exe/ZyPURL.cgi?Dockey=20012ME9.txt

US EPA. 1977. The Report to Congress, Waste Disposal Practices and Their Effects on Ground Water, Sections XI, XIII ("Report to Congress"). nepis.epa.gov/Exe/ZyPURL.cgi?Dockey=2000J6YI.txt.

US EPA. 1990. Federal Financial Demonstrations for Owners and Operators of Class II Oil-and Gas-Related Injection Wells. Document No. EPA 570/9-90-003, Washington, DC. http://www.epa.gov/ogwdw/uic/pdfs/guidance/guide-memo_guidance-67_own-op_hndbk_fin_resp_class2_1990.pdf.

US EPA. 1992. Management and Monitoring Requirements for Class II Wells in Temporary Abandoned Status. UIC Program Guidance #78. Washington, DC. http://www.epa.gov/ogwdw/uic/pdfs/guidance/guide-memo_guidance-78_mgmt_mon_class2_temp_aband_1992.pdf.

US EPA. 2005. Continued EPA Leadership Will Support State Needs for Information and Guidance of RCRA Financial Assurance. Government Print Office: Washington, DC. http://www.epa.gov/oig/reports/2005/20050926-2005-P-00026.pdf.

US EPA. 2011. Plan EJ 2014: Considering Environmental Justice in Permitting. September. http://www.epa.gov/environmentaljustice/resources/policy/plan-ej-2014/plan-ej-permitting-2011-09.pdf.

U.S. Government Accountability Office (GAO). 2005. Environmental Liabilities: EPA Should Do More to Ensure that Liable Parties Meet Their Cleanup Obligations. Government Print Office: Washington, DC.

Appendix A
Pathways of Contamination and UIC Requirements Designed to Mitigate Risks to USDWs

Pathways of Contamination and UIC Requirements Designed to Mitigate Risks to USDWs

The fundamental purpose of the UIC Program is to prevent the contamination of current and potential underground sources of drinking water (USDWs) by keeping injected fluids within the injection well and the intended injection zone. There are six major pathways by which injected fluids can migrate into USDWs, as follows:

1. Migration of fluids through a faulty injection well casing;

2. Migration of fluids through the annulus located between the casing and well bore;

3. Migration of fluids from an injection zone through the confining strata;

4. Vertical migration of fluids through improperly abandoned and improperly completed wells;

5. Lateral migration of fluids from within an injection zone into a protected portion of that stratum; and

6. Direct injection of fluids into or above an Underground Source of Drinking Water.

More detail about each pathway and the major technical UIC requirements developed to mitigate the associated risks to USDWs are provided below.

Pathway 1 – Migration of Fluids Through a Faulty Injection Well Casing

Injection well casing serves multiple functions. It supports the well bore to prevent collapse of the hole and resultant loss of the well; serves as the conduit for injected fluids from the land surface to the intended injection zone; and supports other components of the well. If a well casing is defective or compromised, injected fluids may leak through it, potentially resulting in USDW endangerment.[1,2] To prevent migration of fluids through the casing, well casing should be sufficient to prevent the movement of fluids into any USDWs.

UIC regulations require injection well owners or operators to comply with specific operational requirements designed to minimize migration of fluids through the casing. Foremost among these are the requirements to demonstrate and maintain mechanical integrity (40 CFR 146.8). A

---

[1] US EPA. January 1977. The Report to Congress, Waste Disposal Practices and Their Effects on Ground Water, Sections XI, XIII ("Report to Congress").

[2] US EPA. December 1977. An Introduction to the Technology of Subsurface Wastewater Injection. Chapter 7 ("Subsurface Wastewater Injection").

MIT is used to verify mechanical integrity of the well and confirm the absence of significant leaks.[3,4]

Well integrity can be demonstrated by testing for the absence of significant leaks in the casing, tubing, or packer and the absence of significant fluid movement into USDWs. The regulations, at 40 CFR 146.8, afford owners or operators and Directors options of tests that may be used to detect leaks and fluid movement.

A second protective feature of the UIC Program regulations is that injection wells are constructed with tubing and packer, fluid seal or an approved alternative. Tubing and packer well construction is employed to isolate the casing of the well from injected fluids. Preventing contact between casing and injected fluids reduces the potential for movement of fluids through leaks in the casing and into USDWs.

Pathway 2 – Migration of Fluids Through the Annulus Located Between the Casing and the Well Bore

A second potential pathway by which contaminants can reach USDWs is the upward migration of fluids through the annulus.[5] Under usual injection conditions, injected fluids leave the injection well and enter a stratum that allows the entry of the fluids to varying degrees.[6] Because fluids tend to take the path of least resistance, unless properly contained, they may travel through the wellbore annulus. If sufficient injection pressure exists, the injected fluids could flow into an overlying or underlying USDW.

Measures for the prevention of fluid migration through the annulus (Pathway 2) are the same as those discussed previously for Pathway 1 mitigation. Injection well owners or operators must demonstrate to the satisfaction of the UIC Program Director that there is no significant fluid movement into or between USDWs through the annulus. MITs must be conducted to confirm well integrity and the absence of fluid movement (40 CFR 146.8).

Pathway 3 – Migration of Fluids from an Injection Zone Through the Confining Strata

The third migration pathway the UIC requirements are designed to prevent is fluid migration from the injection zone, through the confining zone, into overlying or underlying USDWs. Upon entry into an injection zone, fluids injected under pressure typically travel away from the well laterally into the receiving formation. In limited situations, if the confining stratum which separates the injection zone from an overlying or underlying USDW is either fractured or permeable, the fluids may migrate out of the receiving formation and into USDWs.

---

[3] See requirements at 40 CFR 146.8.
[4] Geraghty and Miller, Inc. April 30, 1980. Mechanical Integrity Testing of Injection Wells.
[5] The space between the drilled hole/borehole and the injection well casing.
[6] Resistance results from friction created by extremely small openings (pores) in the materials which comprise the injection zone.

The UIC regulations include site characterization, site selection, operation and permitting requirements to prevent fluid migration into USDWs through the confining zone. The regulations require owners or operators to collect and submit comprehensive, site- and project-specific data including information on the geologic characteristics of the injection zone and confining zone(s) to the UIC Program Director for review prior to permit issuance (40 CFR 146.14(a)(l), 146.24(a)(l), 146.34(a)(l)). Historical data may assist EPA UIC permit writers in evaluating an injection well site. An injection well permit should only be issued upon the EPA UIC permit writer's finding that the injection zone is appropriate to receive and retain the injectate and that the confining zone(s) are appropriately characterized and sufficient to contain fluids in the injection zone.

The regulations require that well injection pressure be controlled to prevent opening fractures in the confining strata or otherwise cause the rise of fluids out of the injection zone and into USDWs (40 CFR 146.23(a)). These requirements afford the UIC Program Director discretion to establish injection pressures appropriate for the injection operation.

Pathway 4 – Vertical Migration of Fluids Through Improperly Abandoned and Improperly Completed Wells

UIC site characterization and permitting requirements are designed to mitigate risks associated with fluid migration through improperly abandoned and improperly completed wells into USDWs (Pathway 4). Such migration could occur if fluids move laterally within an injection zone, encounter improperly abandoned or completed wells and flow upward within the well into an overlying USDW or reach the surface. Due to the large number of wells drilled in the past and limitations on historical records, mitigation of fluid movement through this pathway is critical.

To prevent fluid migration through improperly abandoned or improperly plugged wells into USDWs, the regulations require owners or operators to delineate an AoR for each injection well or operation and to identify and locate all wells within the AoR and correct any problems related to improperly abandoned or improperly completed wells before commencing injection.

Pathway 5 – Lateral Migration of Fluids from Within an Injection Zone into a Protected Portion of that Stratum

In most geologic settings and injection scenarios, the injection zone of a particular injection operation will be physically segregated from USDWs by an impermeable confining zone or a series of formations. However, there may be limited circumstances where injection well owners or operators may inject into a non-USDW (a formation not afforded SDWA protection) which is laterally connected to, or proximal to, a USDW. In such situations there may be no impermeable layer or other barrier present to prevent fluid migration into USDWs (Pathway 5).

Injection into non-USDW formations that are laterally connected to USDWs may be permitted depending upon the geologic setting and operational conditions. In such situations, the owner or operator and the EPA UIC permit writer must carefully evaluate the site characterization, well construction and proposed well operation data when establishing permit conditions to ensure that the injectate remains in the injection zone and does not migrate laterally into USDWs. The UIC regulations afford the UIC Program Director discretion to establish appropriate permit conditions on a project-specific basis to ensure USDW protection.

Pathway 6 – Direct Injection of Fluids into or above an Underground Source of Drinking Water

The final pathway mitigated by specific UIC injection well requirements is that of direct injection of fluids into or above a USDW. Such injection presents an immediate risk to public health because it can directly degrade groundwater, especially if the injected fluids do not benefit from any natural attenuation from contact with soil, as they might during movement through an aquifer or separating stratum. To address these concerns, the UIC Class II regulations prohibit injection of contaminants directly into USDWs and permit conditions are established to safeguard USDWs when injection zones are located at shallower zones.

Appendix B
Methods for Calculating the Area of Review

Methods for Calculating the Area of Review

Method Selection

The UIC regulations at 40 CFR 146.6 provide for two approaches to delineating the area of review (AoR): a mathematical approach for calculating a ZEI and a fixed-radius approach.[1] When choosing which approach to require for wells that will use diesel fuels for HF, EPA UIC permit writers should consider that the purpose of delineating the AoR is to identify the area throughout which the owner or operator must search for conduits, such as abandoned wells, that could enable fluids containing diesel fuels to migrate from the injection zone into a USDW.[2]

Calculating the Zone of Endangering Influence (ZEI)

The ZEI is the lateral area in which the pressures in the injection zone may cause injection or formation fluid to migrate into a USDW. In the case of area permits, the ZEI is the project area plus a circumscribing area in which the pressures in the injection zone may cause injection or formation fluid to migrate into a USDW.

The UIC regulations at 40 CFR 146.6(a)(2) provide a formula, known as the modified Theis equation, as an example for calculating the ZEI for a vertical well, pumping over time, in an injection zone. A HF operation creates, within a very-low permeability geologic stratum, a localized, high-density network of interconnected fractures that is very capable of transporting the HF fluids generally consisting of water with a diesel-fuel component. This system may be considered as a porous and confined injection zone and can serve to illustrate why use of the modified Theis equation for calculating ZEIs for long lateral well completions used in HF is problematic. Any application of the modified Theis equation requires that the well-test scenario meets several radial-flow assumptions. Specific vertical-well scenarios may not fully meet all those assumptions, but horizontal, or directionally completed, HF well scenarios significantly violate the following three Theis assumptions:

1. *The injection well penetrates the entire thickness of the injection zone:* While the <u>vertical</u> measurement of the directional completion in a diesel fuels HF application is measured in tens of feet, the vertical thickness of the hydraulically fractured zone is generally several hundreds of feet. Therefore, the directional completion does not approximate a well that fully penetrates the injection zone.

---

[1] Fracture lengths shown in Figures 1 – 5 are for illustrative purposes only.

[2] The Director may ask the owner or operator to apply the fixed-radius approach and if that result is not sufficiently protective, the Director may ask the owner or operator to apply the ZEI approach (or vice versa) to determine if it provides more protection. The Director has the discretion to ask that the approach that is more protective be used. As an example: a ¼ mile fixed radius is applied and the AoR boundary intersects the edge of a drinking water protection area, or is sufficiently close to a public water supply source that the Director considers that HF activities might contaminate a USDW. The Director could ask for application of the ZEI approach. The director could then ask that the approach that provides the more protective AoR be selected.

---

2. *The injection zone is of infinite areal extent*: In the use of diesel fuels for HF application, the injection zone is of limited areal extent within a very low permeability geologic stratum.

3. *The trace of the well onto the land surface is infinitesimal:* In a diesel fuels HF application, the trace of a horizontal or directionally drilled well onto the land surface is not small; rather, it is a line of significant length.

Because the modified Theis equation leads to significant errors if used to calculate the ZEI for horizontal completions, the EPA does not recommend its use in those circumstances. The EPA UIC permit writer may instead consider mathematical models, supported by sufficient field data, to be appropriate to apply to the specific geologic setting for the purpose of calculation of the ZEI. The use of mathematical models often requires a significant body of data.

Using the Fixed One-Quarter (¼) Mile Radius

The second approach for conducting the AoR delineation provided in 40 CFR 146.6 is to use a fixed radius methodology. The owner or operator may use a fixed radius of at least one-quarter (¼) mile around the well bore as the AoR instead of calculating the ZEI, with the approval of the UIC Program Director. The fixed radius is most readily applied to vertical wells.

However, for non-vertical wells, it is necessary to account for the directional portion of the well in order to adequately protect USDWs. For these settings, the EPA has developed the four options below to adapt the fixed one-quarter (¼) mile radius. The permit writer is reminded that, in the case of wells deeper than about 2,000 feet, the extent of induced fractures is greater in the horizontal direction than in the vertical direction, an important factor to consider when applying setback distances from the termination points of fractures or assuring that total fracture extent is included within the AoR. The UIC Program Director, as authorized in 40 CFR 146.6, may require that the AoR be bounded by any of the following:

1. The trace on the land surface of the circumference of a sphere drawn around the directional completion of the well, where the sphere is centered at the mid-point of the directional completion, fully contains all hydraulically induced fractures and has a radius of no less than ¼ mile (Figure 1). (Note: fractures generally do not extend from the endpoints of a directional completion.)

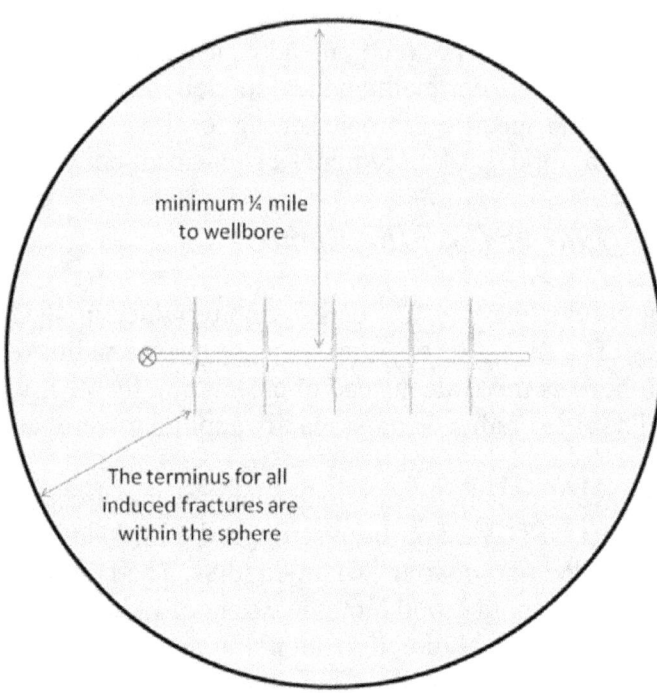

Figure 1. AoR for the trace on the land surface of the circumference of a sphere drawn around the directional completion of the well, where the sphere is centered at the mid-point of the directional completion, fully contains all hydraulically induced fractures and has a radius of no less than one-quarter mile. (Note: Features are not drawn to scale.)

2. The trace on the land surface of the circumference of a sphere drawn around the directional completion of the well, where the sphere is centered at the mid-point of the directional completion, has a radius such that all fractures are completely contained <u>and the termination points of the fractures</u> are no closer to the sphere's circumference than one-quarter (¼) mile (Figure 2).

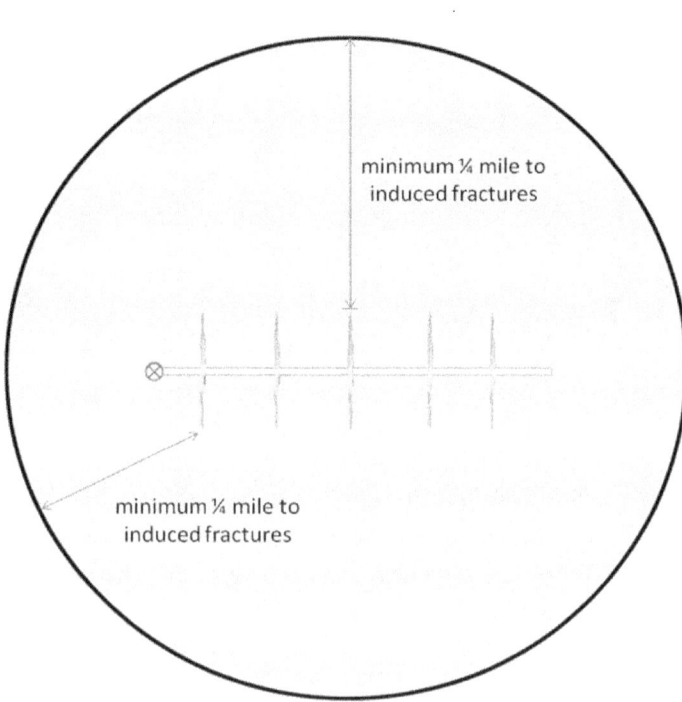

Figure 2. AoR for the trace on the land surface of the circumference of a sphere drawn around the directional completion of the well and centered at the mid-point of the directional completion. The sphere wholly contains all fractures, the termination points of which are no closer to the circumference than one-quarter mile. (Note: Features are not drawn to scale.)

3. The trace on the land surface of the boundary of a cigar-shaped setback from the directional completion, where the cigar shape around the directional completion fully contains all hydraulically induced fractures and has a radius of no less than one-quarter (¼) mile <u>measured from the directional completion </u>(Figure 3). (Note: Increasing the vertical angle of the directional completion reduces the <u>length</u> of the AoR's trace on the land surface.)[3]

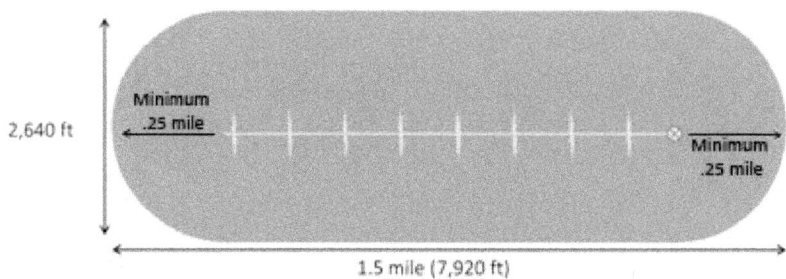

Figure 3. AoR for the trace on the land surface of the boundary of a cigar-shaped setback from the directional completion, where the cigar shape around the directional completion fully contains all hydraulically induced fractures and has a radius of no less than one-quarter (¼) mile <u>measured from the directional completion. The total width of the cigar shape is 2,640 feet.</u> (Note: Features are not drawn to scale.)

---

[3] As the angle of the directional completion approaches vertical, the trace on the land surface approaches a fixed radius around a vertical well.

4.   The trace on the land surface of the boundary of a cigar-shaped setback from the directional completion, where the setback is no less than one-quarter (¼) mile from the estimated end of the fractures. (Note: Increasing the vertical angle of the directional completion reduces the length of the AoR's trace on the land surface.)[4]

Figure 4 below provides an example in which the AoR is defined by the trace on the land surface of a cigar shape drawn one-quarter (¼) mile beyond the endpoints of hydraulically induced fractures that extend 200 feet beyond the directional completion, for a total setback distance of 1,520 feet from the completion (fractures do not extend from the ends of the directional completion.) The completion is horizontal and one mile long. Note that the lateral boundaries of the AoR are curves that are, at their closest point, ¼ mile from the horizontal completion.

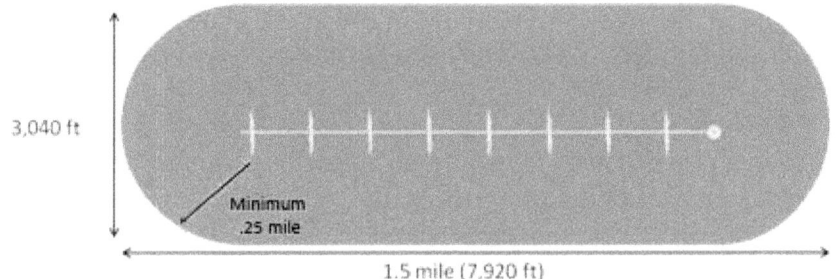

Figure 4. AoR for a cigar-shaped setback drawn ¼-mile beyond the endpoints of 200 feet long induced fractures along the length of a horizontally completed well. The total width of the cigar shape is 3,040 feet. (Fractures do not extend from the endpoints of the directional completion.) (Note: Features are not drawn to scale.)

Multiple horizontal wells are installed at many HF sites. The arrangement of these wells depends on the nature of the hydraulic properties of the zone targeted to undergo HF. Figure 5 presents an AoR that is a composite of the AoRs for three parallel horizontal wells.

---

[4] As the angle of the directional completion approaches vertical, the trace on the land surface approaches a fixed radius around a vertical well.

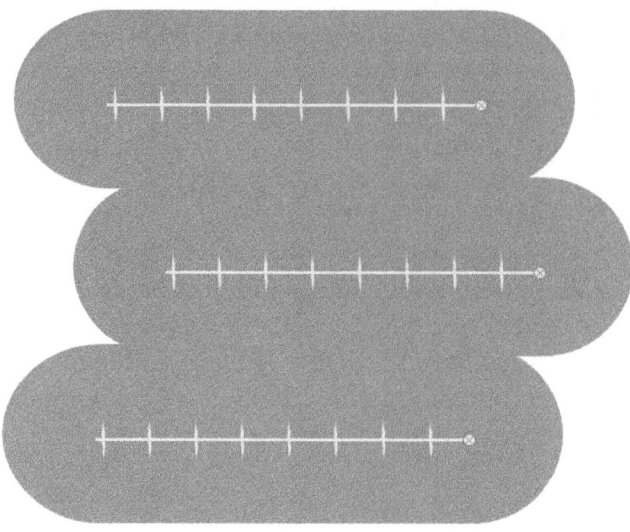

Figure 5. AoR that is a composite of the AoRs for three separate horizontal wells. (Note: Features are not drawn to scale.)

Area Permits. For an area permit, the AoR would be defined by the furthest extent of all well completions (lateral and vertical) plus a circumscribing area, the width of that circumscribing area is either:

1. A fixed difference of (a) at least one-quarter (¼) mile beyond the furthest extent of all well completions and (b) no less than the estimated hydraulically induced fracture length such that all induced fractures are contained within the AoR, or

2. A distance that is calculated by a model according to the criteria set forth in 40 CFR 146.6, but no less than needed to incorporate the farthest extent of fractures emanating from any well covered under the area permit.